Chart

The Mapping the Earth Files

Discovery Channel School
Science Collections

DISCOVERY CHANNEL

SCHOOL™

For information regarding permission, write to Discovery Channel School, 7700 Wisconsin Avenue, Bethesda, MD 20814.
Printed in the USA ISBN: 1-58738-145-1

1 2 3 4 5 6 7 8 9 10 PO 06 05 04 03 02 01

Discovery Communications, Inc., produces high-quality television programming, interactive media, books, films, and consumer products. Discovery Networks, a division of Discovery Communications, Inc., operates and manages Discovery Channel, TLC, Animal Planet, Discovery Health Channel, and Travel Channel.

Writers: Jackie Ball, Monique Peterson, Anna Prokos. **Editor:** Anna Prokos. **Photographs:** p. 2, *Endeavour*, NASA; p. 3, sextant, NASA; p. 4, shuttle crew, NASA; p. 5, NASA (all); p. 6, Landsat 7 satellite, NASA; pp. 6–7, Landsat image of Denver, NASA; p. 8, stick map from Micronesia, Dept. of Anthropology/Smithsonian; p. 8, Psalter map, Granger Collection; p. 9, mosaic map, ©Adam Woolfit/Corbis, double map, Visual Language, geologic map of Ireland, ©CORBIS; p. 10, Ptolemy map, Granger collection; p. 11, John Cabot, Brown Brothers, Ltd.; p. 12, astrolabe, compass (both), PhotoDisc; p. 13, GPS, Comstock, radar, sextant, PhotoDisc; pp. 14–15, NOAA (all); p. 16, Lewis and Clark (both), Brown Brothers, Ltd.; p. 17, journal map of Missouri River, Beinecke Rare Book and Manuscript Library; p. 18, John Harrison, Culver Pictures; p. 19, H-4 and H-1 (both), Granger Collection; 24, comet, NASA, Alvarez, ©Roger Ressmeyer/CORBIS; p. 25, gravity anomaly map, NOAA; p. 26, Radarsat, NOAA; pp. 26–27, glacier, Corel; p. 31, game pieces, PhotoDisc, Fiji, Corel. **Illustrations:** pp. 22–23, sonar vehicle mapping the ocean floor, Mike Saunders; pp. 28–29, road map and topographic map (both), Martin Walz.

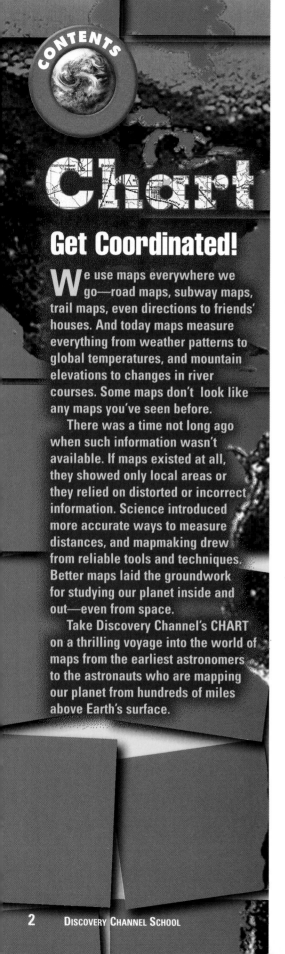

CONTENTS

Chart

Get Coordinated!

We use maps everywhere we go—road maps, subway maps, trail maps, even directions to friends' houses. And today maps measure everything from weather patterns to global temperatures, and mountain elevations to changes in river courses. Some maps don't look like any maps you've seen before.

There was a time not long ago when such information wasn't available. If maps existed at all, they showed only local areas or they relied on distorted or incorrect information. Science introduced more accurate ways to measure distances, and mapmaking drew from reliable tools and techniques. Better maps laid the groundwork for studying our planet inside and out—even from space.

Take Discovery Channel's CHART on a thrilling voyage into the world of maps from the earliest astronomers to the astronauts who are mapping our planet from hundreds of miles above Earth's surface.

The Mapping the Earth Files

A sextant tells you where you are.

See page 13.

Final Project

Earth's orbit, February 12, 2000

Mapmaking reached new heights when the space shuttle *Endeavour* blasted out of our atmosphere. Its mission was to map the entire earth, one piece at a time. Never before could we make a complete and accurate map of the whole planet.

Maps have come a long way since the ancient Babylonians first chiseled one on a clay tablet. In the past some maps were unreliable. Before navigational tools were developed, ancient explorers often used guesswork—and lots of luck—to figure out their location on land or the high seas. Starting in the 1200s compasses and timepieces came into use. They were not as precise as modern mapping tools, making it difficult or impossible to determine latitude and longitude. Careful study of land features and coastlines did lead to some surprisingly accurate maps, however. Skilled mapmakers could draw continents and islands that weren't too far off from contemporary maps. That's quite a feat, especially because they had no way of seeing the world from above.

Once mapmakers had tools and techniques to produce more accurate maps, people began to see the value of showing information in a graphic display. Starting in the 19th century, maps showed rivers, mountain elevations, shifting coastlines, erosion patterns, and other aspects of the constantly changing surface of the earth. Radar, aerial photography, underwater vehicles, space missions, the Internet, and

The shuttle's Radar Topography Mission insignia

other such technological breakthroughs have brought mapmaking to a new level.

The *Endeavour*'s Radar Topography Mission gathered data about plains, valleys, hills, and other details of the planet's surface. During their 11-day mission, *Endeavour*'s six crew members gathered data for most of the earth's surface—enough information to fill 20,600 CDs. It will take scientists approximately two years to analyze the material and turn it into completed maps.

These maps will help scientists study erosion, earthquakes, flooding, volcanoes, landslides, and weather and climate change. Maps will also help scientists predict climatic and geological changes on the planet's surface, and guide governments in making decisions about land areas. All this information propels scientific discovery and leads the way for Earth's explorers of the future.

The crew of the space shuttle *Endeavour*

1. An antenna on the *Endeavour* sends a beam of radar waves, which bounce back to the shuttle (right).

2. Each signal produces a radar image (below).

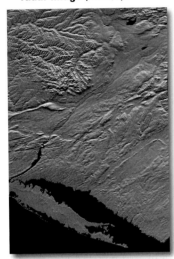

3. A computer combines radar images and adds color to show differences in altitude (below).

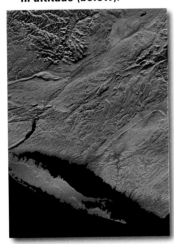

4. Scientists and cartographers use the radar images to create 3-D maps and topographic maps. The 3-D map at right shows southern Connecticut and the Catskill Mountains in New York.

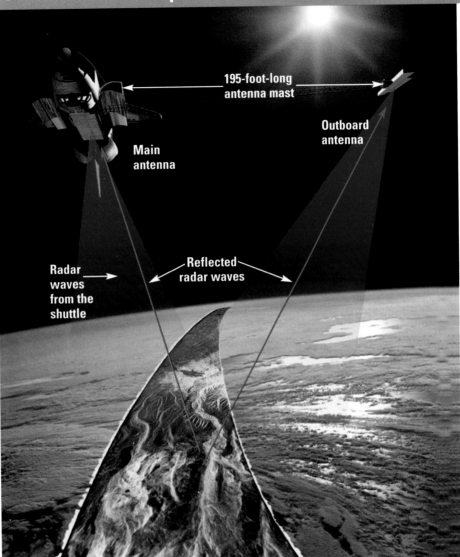

195-foot-long antenna mast

Outboard antenna

Main antenna

Radar waves from the shuttle

Reflected radar waves

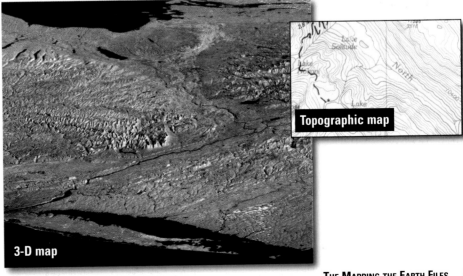

Topographic map

3-D map

Photo Ops

Seeing the world from a satellite's point of view

Background: Landsat photograph of suburban Denver, Colorado

Q: We've been put into special communication with something way over everyone's heads: an American Landsat satellite. Yoo-hoo! Can you hear us up there?
A: Loud and clear, even from 560 miles up—or 700 kilometers, if you prefer. No need to shout and disturb my peace and quiet.

Q: Can you chat for a minute?
A: I guess so, but don't expect me to stop working. I never take a break. Can't—I'm too busy.

Q: Busy doing what?
A: Oh, circling, for one thing. Or make that, for 14 things. American Landsats like me orbit the earth 14 times every day.

Q: Don't you ever get bored going around in circles?
A: Nope. Too much to do. Too much to see. The whole world is down there! How could it be boring? Although we all get a little dizzy from time to time.

Q: What do you mean, "We all"? Who else is up there?
A: Other satellites. Space is loaded with 'em. Must be thousands up here. Almost every country has a bunch of its own. We try to stay out of each other's way, but it's not always easy.

Q: What exactly do all of you do?
A: We're high-flying, super-sophisticated photographers. We take pictures with fancy cameras called remote sensors.

Q: You mean, you do aerial photography? Like photos of fields and stuff taken from planes?
A: Kind of like that, but much more exciting. Those old-fashioned photos are so flat and dull. They're so . . . yesterday. The pictures we take are TOMORROW!

Q: How are your pictures different from the old-fashioned kind?

A: In a word, they're masterpieces. They show every curve and dip and shadow of the earth—every teensy detail. Not just that: My remote sensors are so powerful they record a world beyond what's visible to the naked eye—heat and invisible light. And they record it in gorgeous false color.

Q: Whoa! False color? You mean, as in fake?
A: No, no. False color is just a term. Regular photography records only light visible to the human eye. But my sensors record infrared light: the light beyond the red end of the color spectrum. It's called false color. My photos can show healthy plants as red and diseased ones as blue. Or heavily populated areas in pink and forested ones in brown.

Q: Regular old-fashioned cameras must be green with envy! But what's the point of taking pictures like that? Wouldn't it be better to show things as they are?

A: Wrong, wrong, wrong. Where's the progress in photographing what you can see anyway? With MY pictures, scientists can look ahead. They can monitor problems before they get out of hand. Because infrared light sends out heat, scientists can look at a picture and chart the beginnings of a forest fire, overdevelopment, certain kinds of pollution—almost anything.

Q: That IS pretty exciting.

A: You can say that again. Actually, at the risk of being immodest, satellite photography like mine is the biggest breakthrough in mapping since the invention of longitude and latitude. Wouldn't you agree?

Q: The invention of what?

A: Longitude and latitude. Those are the imaginary lines geographers and mapmakers use to measure off the world. They give everything on Earth a specific location, and that's a good thing.

Q: Why?

A: It makes every place easy to find, no matter where you're coming from. Longitude and latitude measurements are used with my pictures, so scientists know exactly where the problem or situation is taking place. Otherwise there's just too much guesswork.

Q: Oh, that makes sense. But what's the difference between longitude and latitude?

A: Think long for longitude. Vertical. Up and down. Top to bottom. North-south. Another name for a line of longitude is a meridian. Meridians divide up the world into vertical slices, like an apple cut into wedges.

Q: So latitude goes around?

A: Correct! Lines of latitude go east-west, circling the world parallel to the Equator. That's why another name for a line of latitude is a parallel. Parallels divide the world into horizontal slices, like an apple cut into rings.

Q: But how are longitude and latitude used to measure the world? Where do they start?

A: They start in the same place any number system starts—with zero. Mapmakers defined the line of longitude running through Greenwich, England, as being zero degrees longitude. That line is called the prime meridian. Every other place on Earth measures as being a certain number of degrees east or west of the prime meridian.

Q: How about latitude?

A: That starts with zero, too. But the zero is at the Equator. Every other place on Earth has a latitude that's so many degrees north or south of the Equator.

Q: Interesting. Sounds very . . . precise.

A: That's the point exactly— precision! Every place on Earth has an absolute location. No matter where you're starting out, you can take aim at your destination and then use other tools to figure out how to get there.

Q: But what about you? I mean, you're hundreds of miles in the sky. Sure, longitude and latitude are important down here, but I'm curious how you got to be where you are.

A: I've been up here since April 1997. NASA and the U.S. Geological Survey run my operation jointly. They're the ones who use all the data I send them every day, and they make it available to scientists around the world.

Q: Won't you get tired after a while?

A: Oh, sure. And maybe just a wee bit outdated. They keep coming up with new, improved versions. The first Landsats were launched in 1972. I'm the seventh in a long, illustrious line. One day I'm sure there'll be a Landsat 8.

Q: Well, for someone so far above us all, you've certainly brought mapping down to Earth. I hope you'll stick around for a while.

A: Thanks. And that's the right word: around. And around, and around, and around . . .

Activity

LANDSAT ROUNDUP The American Landsat program has come a long way since the first satellite took orbit in 1972. Collect some information on the program from the Internet and library sources and put together a timeline tracing major developments. What are the milestones for the mission? How has the mission changed?

Mapping the World Over

Compare the maps on these pages to see some different world views over time.

SMALL WONDER

This map is only about 6 inches across, but it includes lots of information. It appeared in a Book of Psalms published in 1250, and it's the oldest surviving map showing biblical events: Moses is crossing the Red Sea at the top right. Instead of north, east is at the top of the map, but if you rotate it so that north is up, you might start to recognize some places. Which areas can you identify? What is at the map's center?

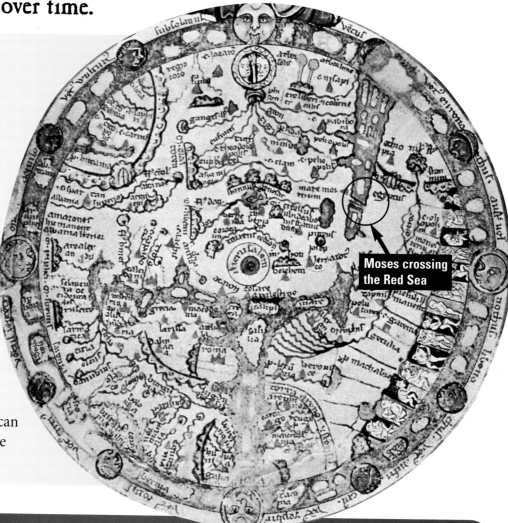

Moses crossing the Red Sea

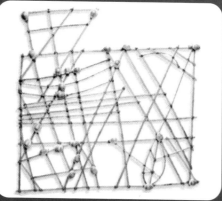

STICK AROUND

Ancient Polynesians navigated the Pacific starting 2,000 years ago, long before people from Europe even knew this ocean existed. The navigators studied ocean currents and noted the locations of thousands of islands. Ship captains aiming for a small island would be in serious trouble if they didn't know how the currents flowed. Polynesian sailors recorded this information on charts (see example at left), using shells to represent islands and sticks to indicate the currents and wave patterns surrounding them.

East Is Up

The oldest existing map of the Holy Land, made of mosaic tiles, was once the floor of a 6th-century church. In this map, west is at the top and north is to the right. The body of water is the Mediterranean Sea, and the river to the left is the Nile. You can tell right away this map isn't drawn to scale: The city of Jerusalem appears almost as big as the Mediterranean Sea! Roman surveyors and road builders used the column at the city gates as a reference point when measuring distances between Jerusalem and other locations.

Nile River

Mediterranean Sea

City gates of Jerusalem

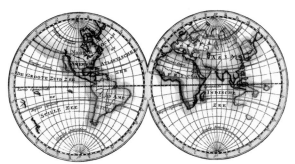

Seeing Double

During the 1700s scientific advances became a driving force in world maps. The map above shows a double hemisphere, a method of mapping the world that began in the 17th century. Showing two hemispheres featured the difference between the Old World (Asia and Europe) and the New World (North and South America), the area that attracted exploration and settlement in the 1500s and 1600s.

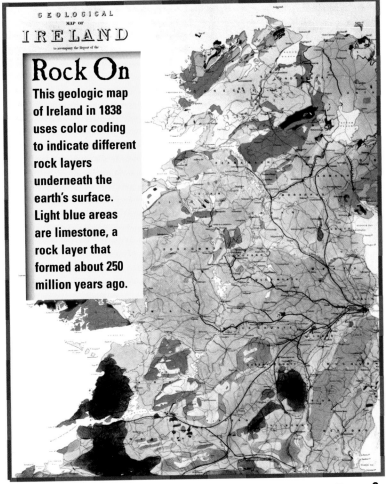

GEOLOGICAL MAP OF IRELAND

Rock On

This geologic map of Ireland in 1838 uses color coding to indicate different rock layers underneath the earth's surface. Light blue areas are limestone, a rock layer that formed about 250 million years ago.

SCRAPBOOK

"Where Are We?"

Before sophisticated navigational tools and techniques existed, explorers had a tough time finding their way.

At first explorers relied on maps made by mapmakers who rarely went exploring themselves. Sometimes these maps included superstitions and legends, leading confused explorers on many a wild goose chase. Once the explorers started making their own maps in the 1500s, the story changed for the better.

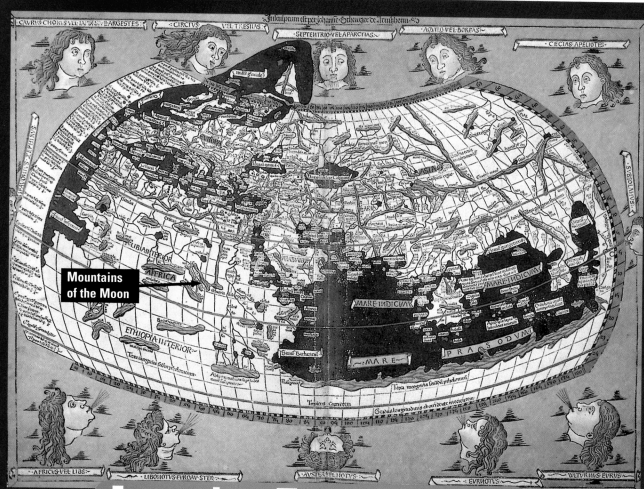

Mountains of the Moon

Learning From Mistakes

Ptolemy (TOL uh mee), a Greek astronomer during the second century AD, made maps of the world that were considered accurate for centuries—but they contained colossal errors. One map showed a nonexistent sea separating Europe and Asia.

Ptolemy's maps showed that the source of the Nile River was a mountain range called Mountains of the Moon. Well into the 19th century explorers searched the continent of Africa for these mountains, all without success.

Getting Oriented

Explorer John Cabot attempted to reach Asia by traveling a northerly route across the Atlantic Ocean. Cabot reached the Gulf of St. Lawrence, near present-day Newfoundland, Canada. He returned to England and brought his son, Sebastian, on his next voyage in 1498. John Cabot didn't survive that journey, and because no logbook or maps survived, no one knows what became of him. Sebastian did return, however, and boasted of both voyages, claiming to have discovered a new route to Asia. Long after Sebastian's death, expert cartographers and explorers investigated his story and concluded that John Cabot, not Sebastian, deserved the credit for first sighting Newfoundland.

Oops

Seeing Isn't Always Believing

Explorers used guesswork and observation to understand uncharted territory. Occasionally their eyes deceived them. The first Europeans to sail along Baja (BAH-ha) California thought it was a large island. In 1539, Spanish explorer Francisco de Ulloa figured out he was sailing in a gulf, later named the Gulf of California, and that Baja was connected to the North American mainland.

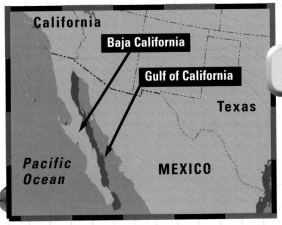

California
Baja California
Gulf of California
Texas
Pacific Ocean
MEXICO

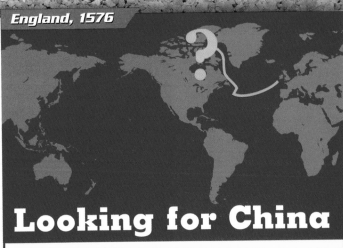

Looking for China

Some explorers didn't realize how big North America was. English soldier Sir Humphrey Gilbert was certain that North America was an island rather than a continent thousands of miles wide. He believed that if sailors could locate the island's northern coast, they could follow it to navigate to China. An enterprising merchant could use such a route to find riches, as Gilbert wrote:

> This discovery hath reserved for some noble prince or worthie man, to make himself rich, and the new world happie.

Queen Elizabeth I funded explorer Martin Frobisher to find this passage. Frobisher quickly found that North America extended much farther north than Gilbert supposed. And the farther Frobisher and his crew sailed, the colder it got.

When sailing along a northern coast didn't work out, they looked for a route through an inland waterway. Explorers went up rivers and inlets searching for what they called the Northwest Passage. We now know that no such passage through North America exists.

Activity

SURE LINES Identify a specific peninsula or bay and find it on a map and a globe. Describe the coastlines from above, from a bird's-eye view. Examples could include the Baja California Peninsula, the Delmarva Peninsula, the Hudson Bay, and the Chesapeake Bay. How could the peninsulas be mistaken for islands? Could the bays be mistaken for oceans? As a navigator, how could you plot your course to correct a mistake?

TOOLS OF

You may have heard the expression "lost without a compass." But have you considered what it's really like to be in the middle of nowhere, with no way of telling what direction you're heading or which direction you've come from? Navigators and explorers faced this problem in the earliest days of sea voyaging up to the Middle Ages. See how they worked around it, and how the necessary tools came into being through the scientific method.

200 BC
HOW FAR TO GO?
Ancient Greek mapmaker Eratosthenes calculates Earth's circumference on the longest day of the year. Measuring angles of the Sun's rays in two different locations, Eratosthenes concludes that Earth's circumference is 28,566 miles (46,000 km). That's close to the measurement used today: 24,840 miles (40,000 km).

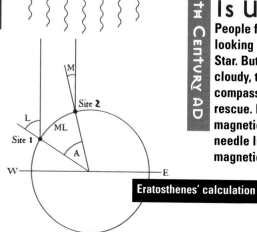

Eratosthenes' calculation

13TH CENTURY AD
WHICH WAY IS UP?
People figure direction by looking at the Sun or North Star. But when the sky is cloudy, the magnetic compass comes to the rescue. Earth has a strong magnetic field, so the compass needle lines up with the magnetic North Pole.

Magnetic compass

15TH CENTURY
LATITUDE ATTITUDE
The invention of the astrolabe in 1484 helps navigators determine their position. This device measures the angle formed by the horizon and the Sun or another star, which gives the latitude, or a north-south position. The astrolabe also helps mapmakers mark latitude lines on a map.

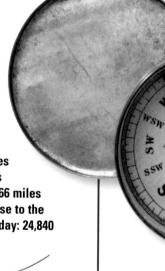

Astrolabe

THE TRADE

18th Century

Pinpointing It

Invented in 1731, the sextant allows explorers to measure the altitude of the Sun or any other celestial body, from which they can determine a geographic location. In the 1760s John Harrison invents a seagoing clock; it calculates east-west positions, or longitude. (Read more about it on pages 18–19.) Explorer James Cook uses it to make detailed charts of a voyage. Mapmakers produce accurate navigational charts based on Cook's measurements.

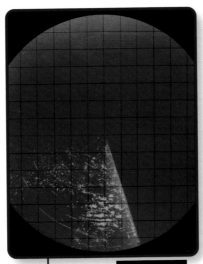

Radar image

20th Century

Radar Revolution

When radar is developed in 1935, scientists can locate objects without seeing them. Radar projects radio waves against an object, for example a distant ship or submarine, to determine its position in space, its size and shape, and the speed and direction it is moving.

2000 and Beyond

Spaced-Out Navigation

The U.S. Department of Defense develops the Global Positioning System (GPS), which can give the location of any object on Earth within about 60 feet. This space-based system makes use of 24 satellites that provide an accurate position regardless of weather.

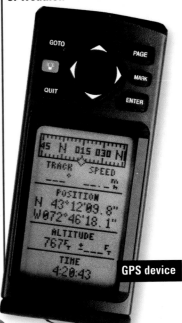

GPS device

Sextant

Activity

GET LOST! Select two leaders of the class to set up a compass course. The leaders will plot out seven spots around your schoolyard or nearby park, setting out directions from one spot to the next. They will provide a full list to three class groups. Each list will have the direction and number of paces to the next spot. Each group starts at a different spot and at the signal tries to complete the seven-spot course as quickly as possible. How big is a pace? Who knows how to read a compass?

Weather Or Not

Thanks to satellites, weather mapping is becoming more precise. Weather maps provide a forecast for an area and give scientists enough information to predict climate changes on a global scale. Check out how weather maps tell the real story— both the present and the future—in these three cases.

EYE OF THE STORM

Hurricanes do damage with high winds and extensive rainfall. This leads to flooding, which can mean disaster for people living in the region. Hurricane Mitch devastated rural areas of Central America in October 1998, taking thousands of lives.

Satellites can keep an eye on dangerous rainfall levels by measuring different types of energy in storm clouds. Infrared, or visible light, readings during storms match up to rainfall measurements on the ground. This could be a lifesaver in underdeveloped areas. Satellite maps can help warn residents in rural communities of flash floods so they can coordinate evacuation efforts.

A satellite image shows the areas of heaviest rainfall from Hurricane Mitch in October 1998.

Rainfall amounts per hour

48-56 inches

24-48 inches

16-24 inches

4-16 inches

Over Exposure

The earth's ozone is the only atmospheric gas that can absorb large quantities of ultraviolet radiation from the Sun, so it protects the planet from deadly solar rays. Scientists study Earth's ozone layer with the Total Ozone Mapping Spectrometer, or TOMS, which is hooked up to a satellite. Computers use TOMS data and color-coding to map ozone concentrations around the world.

The maps of the South Pole at right show a hole in the ozone layer above Antarctica and how it has changed over time. Compare the maps with the scale. What do the color patterns since October 1979 tell you about the ozone hole over Antarctica?

October '79

October '81

October '82

October '85

October '88

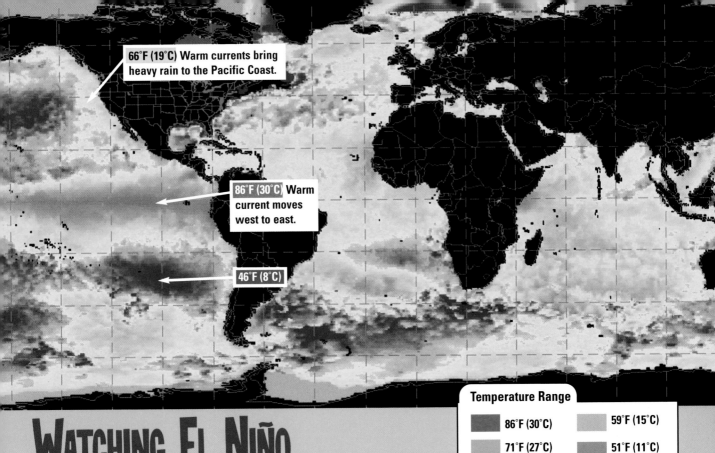

66°F (19°C) Warm currents bring heavy rain to the Pacific Coast.

86°F (30°C) Warm current moves west to east.

46°F (8°C)

Temperature Range

86°F (30°C)	59°F (15°C)
71°F (27°C)	51°F (11°C)
66°F (19°C)	46°F (8°C)

WATCHING EL NIÑO

When water temperatures in the Pacific Ocean change drastically, scientists know El Niño is on its way. This weather event occurs about every seven years, starting as a current of warm water off the west coast of South America. It brings above-average rainfall and severe storms, winds, and flooding to countries around the world.

The U.S. Climate Diagnostics Center uses weather maps to get a handle on when El Niño is next approaching. Higher temperatures in the Midwest, Alaska, and Canada and below-average temperatures in the southeastern U.S. are an early indication. The weather map above, made in 1998, illustrates how the current spreads in the Pacific Ocean. Satellites create such maps using multi-spectral scanning, or MSS, which picks up radiation data to show different temperatures on the earth's surface. Weather maps show this data in patterns of color.

The colors on this scale represent the ozone's different levels of thickness: Purple and violet indicate greater UV exposure.

Dobson Spectrometer scale

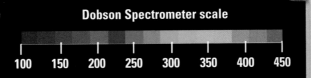

100 150 200 250 300 350 400 450

October '90   October '92   October '94   October '96   October '97

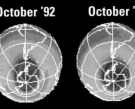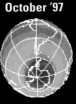

Activity

FOLLOW THE FRONTS Collect weather maps from *USA Today* for one week. What information is on all of the maps? Is there information you think should be added to the map to make more sense? Is there a connection between front lines and mountains? Between front lines and oceans? Would the map make more sense if mountains and rivers were not included?

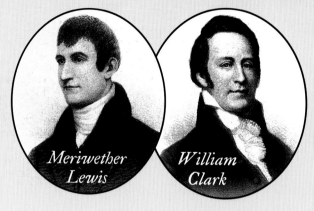

Meriwether Lewis

William Clark

St. Louis, Missouri, May 21, 1804

> The object of your mission is to explore the Missouri River . . . by its course of communication with waters of the Pacific Ocean . . . The courses of the river . . . may be supplied by the compass, the log-line & by time, corrected by the observations themselves.
>
> — *Thomas Jefferson*

With these notes, President Thomas Jefferson instructed Meriwether Lewis and William Clark to set off on one of the most remarkable adventures in American history. No roads or reliable maps existed. The explorers had only the Missouri River and the president's assignment: to follow this river to other waterways until they reached the Pacific. Lewis and Clark—and their team of 28 men—embarked on their famous expedition on May 21, 1804.

Finding their way was no easy task. Their only navigational tools were a sextant, an octant, and the book Nautical Almanac, which gave latitude readings. Accurate longitude readings did not yet exist for this part of the globe, so Lewis and Clark used the location of the stars, Sun, and Moon to estimate their east-west position. Most of the journey was in boats, traveling up rivers. The explorers drew maps in their journals, showing the physical features that they observed, such as rivers and mountains, and plotting these to whatever longitude and latitude their tools and techniques indicated.

CRITICAL CROSSROADS

Because they didn't have maps to guide them, the team often came to a point where they weren't sure which way to go. Travelling up the Missouri River on June 3, 1805, they reached a fork, where the river divided into two branches. They needed to stay on the Missouri to reach another waterway, the Columbia River. If they followed the wrong branch and headed deep into the Rocky Mountains, they'd be in real trouble. Lewis wrote in his journal:

> *To mistake the stream . . . and to ascend such stream to the Rocky Mountains or perhaps much farther before [knowing] whether it did approach the Columbia or not, and then be obliged to return and take the other stream . . . would probably so dishearten the party that it might defeat the expedition altogether.*

EXPLORING OPTIONS

They sent parties up both forks to gather information, gaining high ground so they could see where the rivers flowed. When the groups had returned to camp, Lewis and Clark considered all the information to decide which was the Missouri. Lewis wrote:

> *We took the width of the two rivers, found the left-hand or S. fork 372 yards, and the N. fork 200. The north fork is deeper than the other, but its current not so swift . . . In short, the air and character of this river is so precisely that of the Missouri below that the party with very few exceptions have already pronounced the N. fork to be the Missouri.*

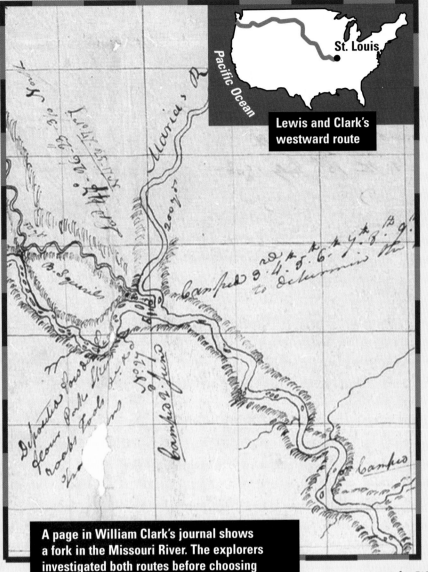

As the explorers had done at every point of their journey, Lewis studied his physical surroundings and made notes on how they related to the coordinates his navigational tools indicated. He had his answer. "I now became well convinced that this branch of the Missouri had its direction too much to the north for our route to the Pacific."

Lewis and Clark were correct: The south fork was the Missouri, and the north fork was a Missouri tributary they named Marias River. They reached the Columbia River in October, safely on the far side of the Rockies before winter set in.

A month later they reached the Pacific Coast, at the mouth of the Columbia. They had spent the past year and a half in the wilderness, moving steadily toward their goal without horses or wagons. Along the way they faced many dangers: rugged terrain, wild animals, harsh winter weather, and encounters with hostile Native Americans. But the expedition was more than just a wilderness adventure. When the explorers made their triumphant return to St. Louis on September 23, 1806, they carried hundreds of pages of notes, descriptions, and maps recording what they had seen.

This information all went into the first detailed map of the West. Completed in 1810 by William Clark, the map showed the entire upper Missouri River system and how it connected to the Columbia River basin. Such data would prove invaluable to later generations of Americans who chose to make the difficult journey to start new lives in the West.

A page in William Clark's journal shows a fork in the Missouri River. The explorers investigated both routes before choosing to follow the south fork.

Lewis and Clark disagreed with the others that the north fork was the Missouri. However, they didn't want to go against the group right away. Instead, they discussed the matter and set off the next morning: Lewis to explore the right, or north, fork and Clark the south. Lewis walked along the banks of the stream 60 miles over two days, taking careful note of his position.

The north fork, which I am now ascending, lies to my left ... on its western border, a range of hills about 10 miles long appear to lie parallel with the river, and from hence bear North 60 degrees West.

Activity

WESTWARD (OR EASTWARD) HO! Lewis and Clark traveled without a map. But you don't have to worry about that disadvantage. Get a map of the United States that shows latitude and longitude. Use the map to find the latitude and longitude of your state boundaries. Then use the Internet to find the exact latitude and longitude of your school or your home.

The Long Journey to Longitude

John Harrison's invention was a mapping and navigation breakthrough.

Lost at Sea

Say you're boarding a cruise ship for your summer vacation. What if your captain has no idea where to steer the ship? That was a reality before the 18th century, when it was impossible for a sailor, explorer, or captain to know the east-west position, or longitude, at sea or on land.

The problem? Longitude is most accurately measured by keeping track of time—the time aboard a ship and at the home port. To determine longitude at sea, a navigator observes the height of the Sun. At noon, for example, the Sun is at its highest point; and if the clock says 9 a.m., there's a three-hour difference. Each hour's difference between the ship and home port equals 15 degrees longitude.

It may seem simple, but hundreds of years ago people could not know the exact hour at two different locations. Clocks aboard ships slowed down, sped up, or stopped completely thanks to movement, the temperature, salt, and water. Captains relied on charts, compasses, and lots of luck. Still, ships commonly got lost at sea or ran aground.

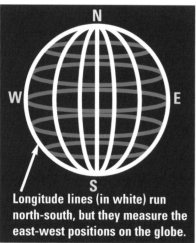

Longitude lines (in white) run north-south, but they measure the east-west positions on the globe.

As more ships set out to explore new lands, unknown longitude became a huge problem. Galileo, Isaac Newton, and Edmond Halley were among the scientists who searched for a solution. Then John Harrison got to work.

Man on a Mission

In 1714 Britain had colonies all over the world, but no way to navigate the oceans accurately. How could they rule the colonies if they couldn't find them? So the British Parliament offered the equivalent of several million U.S. dollars to the person who could determine longitude accurately. John Harrison heeded the call.

John Harrison

Born on March 24, 1693, Harrison taught himself how to read. The turning point in his life came in 1712, when he borrowed a manuscript on the nature of philosophy from Cambridge University. But he didn't just read the book. He copied it word for word. Over the next several years, he scribbled his own theories on the pages. His hard work paid off.

In 1713 Harrison made a pendulum clock. Historians believe he studied, tested, and examined other clocks before making his own. His clock had one thing other clocks didn't have: accuracy. Harrison achieved this by designing his clock to operate at a consistent speed, even in motion and changing temperatures. Even the best clocks lost about one minute each day. Harrison began to build the H-1, the clock that would solve the longitude problem.

The Timekeeper

It took Harrison five years to build the H-1. It didn't need cleaning or oiling, and it didn't rust. No matter how hard he tossed it around, the H-1's moving parts stayed in perfect balance. He brought his invention to George Graham, a highly respected watchmaker. So impressed was Graham by the clock's accuracy that he wrote to the British Parliament, requesting that they test it at sea.

On May 14, 1736, Harrison tested his H-1 aboard the ship Centurion en route to Lisbon, Portugal. The rough voyage made him seasick, but the clock kept on ticking. Returning to England, Harrison's ship captain assumed he was approaching a location on the south coast near Dartmouth. But the H-1 placed the ship more than 60 miles (96 km) west. The captain doubted the H-1, but Harrison's device was correct. On returning to shore, Harrison presented his clock to the government officials and waited for the prize money.

During the next few years, Harrison developed the H-2, the H-3, and in 1759 the H-4—his favorite clock. He kept perfecting his clocks, and he finally received the prize money in 1773. Ever since then, John Harrison has been known as the man who first measured longitude.

The H-4 clock is the forerunner of modern precision watches.

An elaborate internal mechanism helped John Harrison's H-1 clock keep accurate time.

Activity

TEST YOUR LONGITUDE IQ. Apply the longitude formula (15° longitude equals one hour's difference from home port) to these exercises. Locate a globe and a calculator to plot your longitude. Imagine you are sailing in the South Pacific, heading east from Sydney, Australia, along one latitude (north/south position). It's 12:00 noon when you leave, and you're traveling at a rate of 30 miles per hour. What is your longitude position when it is 3:00 p.m. back in Sydney? 5:00 p.m.? 10:00 p.m.? See answers on page 32.

Flat-Out Problem

We know now the world is spherical, yet most maps are flat. The trouble is that a flat map doesn't show the surfaces of a sphere correctly. So what's a cartographer to do? Throughout time scientists, astronomers, artists, and mapmakers have tried showing Earth in two dimensions by using different kinds of projections, or flat maps. Projections try to show the earth's dimensions accurately, yet no single projection is completely accurate. Some projections work better than others. Despite their limitations, every flat map of the round earth serves its purpose: to show a view of the whole world at once.

GLOBE GORES

This map shows the round image on a series of connected pieces called gores. If you could cut out this photo and wrap it around a sphere the right size, the pointed ends would come together at the top and bottom of the sphere (like the North and South Poles on a globe). Attach these ends with tape and voilà! A two-dimensional picture is now in three dimensions.

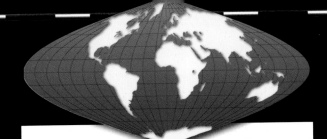

ROBINSON PROJECTION

When a map company asked cartographer Arthur H. Robinson to make a new map, this is what he came up with. The Robinson projection became popular in atlases and wall maps, and in 1990 the National Geographic Society used Robinson's projection in its world atlas. It shows the whole world, but distorts Antarctica into a long, slender continent at the southernmost part of the map.

SINUSOIDAL PROJECTION

A sinusoidal projection shows longitude lines that curve away from a central axis; one longitude line intersects at a right angle with the Equator. The farther out from the central axis, the greater the curve in the longitude and the greater the distortion in landmasses. Notice how North America is stretched almost beyond recognition. As on a globe, all longitude lines meet at the North and South Poles.

MERCATOR PROJECTION

Named after 16th-century Dutch mapmaker Gerardus Mercator, this projection shows longitudes as straight lines that don't intersect. To compensate for this distortion, the distance between latitudes increases the farther north or south they are from the Equator. As a result, the landmasses farther north of the Equator appear much bigger than they are in reality. Note how Greenland appears about the same size as South America. Seagoing ships use Mercator projections because they show continent and island shapes accurately and allow a navigator to plot a course using a straight line.

VAN DER GRINTEN PROJECTION

A projection looks entirely different if the longitudes are straight lines or circular arcs. In 1904 Alphons J. van der Grinten made a map projection with non-intersecting longitude lines. The National Geographic Society was the first to use this projection in 1922 as a basis for its political maps. The projection was widely accepted for many years because it shows the world in a perfect circle and is easy to reproduce on a single page in books.

Activity

A MAP OF YOUR HEAD Blow up a balloon about the size of your head. Cover the balloon with papier-mâché and let it dry. Then draw a picture of your face on the papier mâché. Does it look like you? Now, select one of the projections from these pages. Po the balloon and then carefully cut your paper head to match the projection you have chosen. Lay the paper head flat on a table. Pretty weird, huh? Any flat map is an equal awful distortion of reality. Share your "map" with your classmates, if you dare!

DEEP-SEA
Mission to Map

Hold your breath. You're going down. Way down. To the bottom of the ocean, in fact, to map the ocean floor.

Sonar vehicle

Scanned area of ocean floor

Well, not you—you're part of a national mission to explore the seafloor, and you're aboard the mother ship, which is towing a multi-beam sonar vehicle. The vehicle will send out high-frequency pulses of sound, or pings, to scan the ocean floor. It's sonar (**so**und **na**vigating **r**anging), especially powerful equipment that is critical in making a three-dimensional map of the seafloor.

Wait a minute! Suddenly you're riding on the sonar vehicle itself. You *are* going down! You're being towed underwater, and before you can take a deep breath you hear . . . what's that noise? It's a soft ping, something between a click and a bird chirping. You remember what the captain said at the beginning of your mission: The sonar sends these pings down towards the seafloor.

Duck! Here comes one now! No, the pings won't hurt you, but you have to get out of the way if you want the sonar reading to be as accurate as possible. When the pings hit something, they bounce back to the sonar's hearing device. That means if you're in the way, the sonar might consider you part of the ocean floor and record you instead—and ruin the accuracy of your map.

Faster Means Closer

Ping. This sound bounced back pretty fast, in a matter of seconds, which could mean there's something jutting high up from the ocean floor. The faster a ping comes back, the closer the object it's hitting. Could be a hill. A rock formation. A . . . shark swimming right underneath you?

P-i-n-n-n-g-g-g. That one took longer. Perhaps half a minute or more. It could mean the seafloor is pretty deep at that spot—and free and clear of vents, volcanoes, or ridges. The faster the sound waves return, the shallower the water and the higher the elevation of the seafloor.

You're amazed at the amount of information one little ping can give—so much that for the first time, a complete map of the ocean floor can be made. Explorers have been trying to measure the depth of the ocean for centuries. In

Mother ship

85 BC, the Greek captain Posidonius traveled to the middle of the Mediterranean Sea and tossed two kilometers of rope overboard, with a heavy stone attached to one end. Maybe this method told Posidonius something about the sea's depth there, but it didn't provide much scientific information.

What a difference 2,000 years can make! The sonar vehicle you're riding in is part of the National Oceanic and Atmospheric Association (NOAA), the government agency responsible for mapping the ocean floor. The sonar and your mission are key in providing knowledge about the geologic forces that shape the ocean floor. The images you're creating will give scientists a way to view vast stretches of undersea terrain at a glance. So you're listening carefully: Mud absorbs sound, for example, so a muted ping means a muddy bottom. On the flip side, a strong echo means it's a rocky bottom.

Rugged Terrain

Back on the mother ship, the rest of the team is making sure the onboard computer is calculating the pings correctly. It records the interval that passes between the time the ping leaves the ship and the time it returns from the ocean floor; it uses that amount of time to calculate the depth in each spot. It then takes that data and creates a topographic map, one that shows the hills, valleys, and depths of the seafloor.

Whoa! That ping came back super quick. Something very tall must be right below—maybe a volcano! You imagine hot lava spewing up towards you . . . and suddenly you're back onboard the ship. The crew heard that last ping, too, and they think there is a volcano down there. Once the computer maps this area, you'll be able to see its exact dimensions.

Underwater volcanoes . . . isn't it nice to be back on the ship, where the pings are a little farther away?

Activity

BOUNCE TO THE BOTTOM Sonar works with computers to measure ocean depths. The length of time it takes for the signal to return to the sonar vehicle shows how deep or shallow the ocean is at that point. Pretend you are a sonar computer, and see if you can plot these signals into a cross-section drawing that might resemble the terrain of the ocean floor. Each "1" represents a minute; and each minute represents 5 inches on your sketch: 1.0, .3, .5, .5, . 4, 1, 1.5, 1.2, 1.0, .6, .5, .5, .4, .2, 1.

DINOSAUR DETECTIVE

Maps are a vital clue in one of Earth's most compelling mysteries: What wiped out the dinosaurs 65 million years ago?

An artist's view of an asteroid collision on Earth

Geologist Walter Alvarez used a variety of maps to draw a link between the dinosaurs' extinction and a massive crater in Mexico.

"The history of the earth is recorded in the earth itself. . . . Rocks are the key to Earth history," said Alvarez

With that in mind, in 1991 Alvarez headed to Chicxulub (cheek-soo-LOOB), a site on the coast of the Yucatán Peninsula in Mexico. Alvarez tested rock and clay fragments and found unusually large amounts of the element iridium. Rare on Earth, iridium is found in asteroids. Alvarez concluded that somewhere near that spot an asteroid had struck the earth. If he could find a crater and

match its age to the date of the dinosaur extinction, Alvarez could solve the mystery.

First Alvarez mapped the locations of the iridium sites. To verify the existence of a crater, he consulted a map of Chicxulub that showed irregularities in gravity levels (opposite page).

These maps zoom in on areas of the earth's crust, showing an area's topography and using color codes to indicate different density levels in the surrounding rock. Geologists use such maps to locate mineral deposits underground. The gravity map Alvarez consulted showed areas with the greatest density, where iridium was present. To locate the crater, Alvarez also examined a geologic

Dr. Walter Alvarez, on the left, and his father, Dr. Luis Alvarez, study maps to learn about the earth's past.

map of the area. This type of map not only shows contours and elevations but takes a closer look at different layers of rocks under the surface, indicating which are the oldest and which are the most recent. Geologists study rock layers to link them to different geologic time periods. Using the geologic map, Alvarez proved that the rock layer containing iridium was about 65 million years old.

Surface Clues

In 1991 scientists Kevin Pope, Adriana Ocampo, and Charles Duller plotted the distribution of cenotes (seh NO tays), or sinkholes with a pool at the bottom. They found that the cenotes formed a circle around the Chicxulub site. This circle appeared to offer evidence of an enormous crater buried directly below. No other force could have created such an even circular arrangement of cenotes; the pattern was consistent with the weakening in the earth's crust that occurs at the rim of an impact crater of this size.

The geologic map and the cenote ring gave Alvarez the crater's location. Its center is on the Yucatán coast, but the crater's diameter extends about 125 miles (200 km) across, buried under a layer of rock about half a mile thick on both land and sea. Next he needed evidence of the asteroid's destructive force. "The Chicxulub fireball would be big enough to blow the . . . rock fragments and debris . . . clear through the atmosphere," says Alvarez. "The particles would end at the points all over the world." Studying the site's rock fragments helped him confirm that this had been a major collision. It would have thrown enough debris into the atmosphere to block the Sun's light significantly for several months, killing plant life and all the other species that depended on plants for survival. In short, the collision might have been powerful enough to have caused a mass extinction of life on Earth.

The gravity anomaly map below shows the existence of the Chicxulub crater. Blue indicates areas with the greatest density, where iridium is found.

GULF OF MEXICO

Outer ring of Chicxulub crater

Chicxulub

Cancun

Merida

Campeche

Cenotes

Yucatán Peninsula

MEXICO

Activity

CRATER MANIA Write analogies for the size of a crater. Find five comparisons you could use to explain just how big the crater is. First determine its diameter, then try comparing its size to a state, lake, bay, the area of a city, or another geographic feature. You may want to use fractions; for example, the crater may be "x" times the size of the Great Salt Lake or 1/x the size of the Sahara.

MELTDOWN

IN ANTARCTICA

West Antarctic Ice Shelf

East Antarctica

South Pole

West Antarctica

Amundsen Sea

Satellite photos show what's happening to the ice in Antarctica.

Earth's orbit, 1997

Imagine Washington, D.C., breaking off from the mainland and floating away on the ocean. Now picture it made of ice 1,700 feet (518 m) thick. About 19 square miles (49 sq km) of ice, enough to cover our capital city, breaks off from the West Antarctic ice sheet and floats off to sea every year. If the ice sheet continues to break up at that rate, it could increase sea levels significantly and risk flooding coastal areas around the world.

The situation demands action. In 1997 NASA and the Canadian Space Agency launched

RADARSAT, an orbiting satellite that provides detailed mapping information of the frozen continent. RADARSAT collected data day and night, whether the skies were cloudy or clear. The satellite completed its detailed accurate mapping of Antarctica in just 18 days. Scientists working for the Antarctic Mapping Mission used RADARSAT's images to produce the first high-resolution radar map of Antarctica. This map gives scientists clues to the ice sheet's future—and the fate of global sea levels.

Putting the Pieces Together

The RADARSAT map shows how the West Antarctic ice sheet flows and twists into the sea. Scientists want to learn more about the ice sheet to understand its structure and to find out if it's in danger of collapsing. If the entire ice sheet melts, global sea level will rise by about 17 feet (5 m), flooding coastal cities all over the world.

As scientists constructed the map, they learned some surprising facts. They hadn't known about a complex network of ice streams extending 500 miles (805 km) into East Antarctica. Vast frozen rivers and ice streams flow up to 100 times faster than the ice they channel through. They can cover a distance up to 3,000 feet (914 m) in a year. Scientists now monitor the streams' flow, length, and width, and track changes over time—an impossible feat without a radar map.

Scientists also use seafloor mapping techniques to produce detailed maps of the Amundsen Sea. With these maps, they can trace the glacial history of Antarctica. Colors identify depths: Blue indicates the deepest regions, reds and browns show average depths, and yellows show ridges and other such features.

Mapping Antarctica is not complete. NASA soon will launch a series of satellites to measure the ice sheets covering most of the continent. From this information, scientists will monitor future changes in the ice.

Maps provide a crucial link between past and present. They help us understand global forces that affect our environment. In this case, maps may help prevent a global disaster.

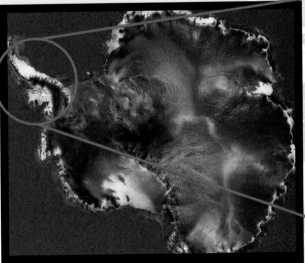

RADARSAT maps show details of Antarctica's geographical features, from its flowing ice streams to breaking ice shelves.

A RADARSAT photograph shows pieces of the West Antarctic ice sheet breaking away to form icebergs.

Activity

SATELLITE LIFE Satellites take a series of pictures over time. Try to find pictures of a community taken over time. Many county offices have such photos; some might be published in a local history book. Look for such a book in your library. How has the area changed over time? How might such photographs be valuable to scientists?

In Search of the SECRET STASH

You're never going to believe this!" Clara Parallax shouted to her friends. "I just found out about buried treasure not far from here!"

"You can't be serious," said Moe Azimuth, always the skeptic.

"Really," insisted Clara. "I was poking around in my granny's old barn, and I found this."

"Let me see," Theodora Gradient said, snatching the paper out of Clara's hand. She read it aloud to the others:

"Wow," said Moe, "Maybe you're right, but shouldn't there be a map with this letter?"

"This was all I found," said Clara. "But I figured we could just use any map of the area. I'm sure it won't make much difference what kind of map we use."

"Hey," Theodora said, her eyes still glued on the old paper. "Things have changed around here since 1875. Isn't the area around Little Elk Creek a new housing development and mall now?"

"Yeah," commented Moe. "My aunt and uncle just

Dear Jessie,   August 5, 1875

No luck finding gold in these parts, so I've gone on to California. I have more silver than I can carry on my horse, so I'm leaving it behind for you. I've buried a couple saddlebags full for you. I've enclosed a map and directions on how to find it. Good luck! Hope to see you west of the Sierras.

Your friend,
Prospector Bill

- Follow Little Elk Creek southwest to Horse Thief Canyon.
- Cross the wooden Pinto Bridge just past Eagle Mountain.
- Head south past Beaver Pond until you reach Rattlesnake Canyon.
- Take the Old Ute Indian Trail southeast through the canyon to Hangman's Tree.
- From the base of Hangman's Tree, walk toward the setting sun between Two Sisters.
- At the southernmost base of the steepest peak, you'll find the entrance to Coyote Cave.
- At the north end of the cave, start digging!

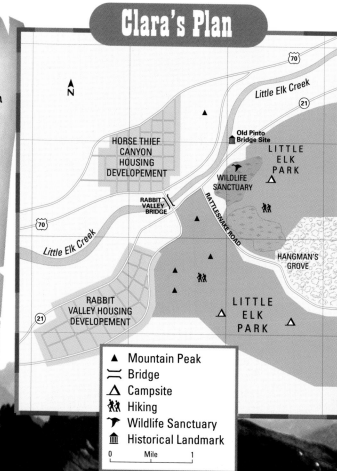

Clara's Plan

- ▲ Mountain Peak
- ⊨ Bridge
- △ Campsite
- 🥾 Hiking
- 🦅 Wildlife Sanctuary
- 🏛 Historical Landmark

0   Mile   1

moved over there. As far as I know, most of the area across the river has been turned into parkland. But I think there are some houses there, too."

"Well, my mom has a road atlas for the whole county," noted Clara. "It includes roads and shows parkland, too, so it's probably our best bet."

"I have some old hiking maps," Theodora chimed in. "They're a little out of date, but they might be helpful."

"Who needs a map when you've got a Global Positioning System receiver?" Moe wondered.

"What's that?" Clara asked.

"Oh, it's a great little electronic gizmo, small enough to fit in your backpack pocket. It can receive satellite signals that give the accurate latitude, longitude, and elevation. There's no way you can get lost with one of those babies. You can keep track of your exact position on the planet, and you can plot your starting point and keep track of the compass direction you're heading in. Plus, the GPS tells your elevation above sea level. I'll ask my dad if I can borrow his. Then I will be your fearless leader to the buried treasure!"

The next day all three friends teamed up with navigational plans in hand. Only one of them had the best information to make it all the way to the buried treasure. Look at the plans and clues below to see who found the silver.

Theodora's Plan

Bridge
Swamp

0   Mile   1

Moe's Plan

1. Locate satellite coordinates from starting point at Little Elk Creek.

2. Follow Prospector Bill's directions and use GPS to keep track of my longitude and latitude.

3. Use GPS as compass to verify bearing.

Clues

Use these clues:

1. The old Pinto Bridge is no longer passable. The nearest crossing is the Rabbit Valley Bridge, about one mile southwest of the old bridge.

2. Beaver Pond has long since become a swamp.

3. Topographical maps include contour lines showing differences in elevation. When contour lines are close together, the slope is steep; when the lines are farther apart, the land is flat. Numbers on contour lines indicate height above sea level.

Answer on page 32

Take My Ship— But Not My Maps!

✪ From ancient times to the 19th century, maps were considered hot property and were very closely guarded. When the Romans stormed a Carthaginian ship, the captain ran his vessel into the rocks. Even though some of his crew drowned, no logs, charts, or maps fell into Roman hands. When the captain eventually reached home, he was celebrated as a national hero.

✪ Spanish explorers in the 16th century weighted their maps with lead. They could throw them overboard if an enemy boarded their ship. The maps would sink to the bottom of the sea without any hope of recovery.

Wacky Map Facts

▸ Before 1593, people in the British Isles didn't always agree on how long a mile was. The Irish mile was 6,720 feet, and the Scottish mile was 5,940 feet. A statute passed in 1593 declared a mile 5,280 feet. Surveyors in 17th-century England measured distances with a foot wheel, an instrument that tracked one foot per revolution. The foot wheel accurately computed that the distance between London and Berwick on the Scottish border was 339 statute miles—not 260 miles, as previously measured in Scottish miles.

▸ Early mapmakers often put east at the top of a map because that's where the Sun rises. Another word for "east" is Orient, so the compass direction at the top of a map is called an orientation.

GLOBAL GAMES

United States, 1906

Most kids couldn't travel around the world in 1906, but they could pretend to take a world trip by playing the Pirate and Traveler Map Game. This was a race around the world. Each player chose a starting point from these four locations: San Francisco, New York, Yokohama, or Paris. Each player followed a specific course laid out on the game board, a map of the world. The player who completed the trip first won. It wasn't just a matter of going the fastest; the game included real-life traveling dangers, like icebergs, sandstorms, and uncertain weather.

London, 1854

Smith Evans developed the Crystal Palace Game to teach people geography. More of a souvenir than a board game as we play today, the game taught kids interesting facts from around the world, from the highest mountains to the pyramids of Egypt to the tigers in India.

Laugh Lines

Q. What do you get when you cross a cowboy with a mapmaker?

A. A cow-tographer.

Q. What do you call a traveler with an ancient map in his hand?

A. Lost.

Q. What did the mapmaker send his sweetheart on Valentine's Day?

A. A dozen compass roses.

SAIL ON

Pacific Ocean, AD 1–1000

You don't have to have high-tech tools to explore unknown waters. Just ask the ancient Polynesians. Long before Western ship captains ventured beyond the Mediterranean Sea, these fearless voyagers were settling islands in an area of the Pacific Ocean covering 10 million square miles (25.9 million sq km). Consider the challenges:

- Vessels were canoes, dug out from thick logs with stone tools and rigged with sails made of coconut leaves.

- There were no compasses or maps: Navigators used the Sun, the stars, and their knowledge of wind patterns and ocean currents to determine direction.

- Canoes were small and narrow, offering only crude shelter—and not much help in ocean storms!

But these sailors made it. The Polynesians had settled most livable islands in the Pacific by the time the first European explorers arrived in the 1600s.

Maps locate places on the surface of the earth. They also show many kinds of information. Depending on the type of map they're making, cartographers will use certain symbols. For example, road maps usually have various colored and shaped lines to distinguish superhighways from smaller roads. Natural resource maps show the crops or minerals found in an area. One thing all maps share is a key that tells what each symbol stands for. It's important to understand those symbols in order to get the intended information from the map.

What does it take to make a map? Here's your challenge:

1. Choose a mapping partner.

2. Decide on a location to map. It can be somewhere in your school, an area of your town, or a nearby park. Choose an area small enough for you to measure and map in detail, but large enough to contain enough information to be interesting. Make sure the area you select allows access. Never trespass on private land.

3. Decide on the type of information your map will present. You may include landforms, points of interest, population, vegetation, or other features. This is also the time to think about scale, such as having 100 feet equal to 1 inch or 100 meters equal to 1 centimeter.

4. Visit your chosen location. Be sure to take along measuring tools, a compass, plenty of paper, and a camera. Keep your scale in mind while collecting measuring tools. And don't forget your pencil!

5. Collect all the measurements you can. Decide how much detail you will include on the map. If you want to mark a large boulder, be prepared to place it accurately on your map. How will you determine and represent the height of a hill?

6. When you make a first draft of your map, consider the types of symbols you'll need to present the information. Does your scale work? Make any adjustments you need to both your map key and the scale. If necessary, use larger chart paper for the next draft. Be prepared to research and possibly visit the location again before you make your final map.

7. Draw your final draft of the map, including the key. Use colors, and make your map as easy as possible to read. Be prepared to explain your entire mapmaking process: why you decided to map that particular area, what you had to do to draw your map, what tools you used, how long it took, what other references you used, how you developed your key, what problems or surprises you faced.

8. Present the map to your classmates. Can they interpret it? Do they recognize the area? Do they have suggestions to improve the map?

Ready for the ultimate challenge? Enter this or any other science project in the Discovery Young Scientist Challenge. Visit *http://school.discovery.com/sciencefaircentral/dysc/index.html* to find out how.

ANSWERS
Heroes Activity, page 19: 3:00 p.m. longitude: 197°; 5:00 p.m. longitude: 227°; 10:00 p.m. longitude: 345°

Solve-It-Yourself Mystery, pages 28–29: Theodora had the best plan to find the buried silver. By using the topographical map, she could get a sense of the physical features of the area. Even though her map was out of date, the geography of the land had stayed the same. Her map showed the steeper part on the Two Sisters butte: She simply looked for the elevation with the tightest ring of contour circles.

Clara's road atlas was more up to date, and it helped her locate many of the landmarks in Prospector Bill's note: She deduced that Rattlesnake Road went through Rattlesnake Canyon and Hangman's Grove was the likely locale of Hangman's Tree. But Clara's map didn't give her enough information about mountain elevations or steepness, so she had no way of identifying the right butte where Prospector Bill had buried the treasure.

Moe's plan turned out to be the worst choice. Modern GPS tools are best used with maps that also show latitude and longitude degrees, like Theodora's topographical map. With nothing more than Prospector Bill's directions, Moe had no way of knowing how the area had changed over time. Without a map, Moe's GPS told him exactly where he was, but not where he needed to go.